General Relativity: Not Exact, But a Useful Approximation

Revised Edition

Albert W. McKinney III
2017 February 17

Copyright © 2017 by Albert W. McKinney III

Library of Congress Cataloging-in-Publication Data
McKinney, Albert William III (1929-)
    General Relativity: Not Exact, But A Useful Approximation
    CreateSpace Independent Publishing Platform,
        North Charleston, South Carolina
    ISBN: 978-1541220959

DEDICATION

This book is dedicated to William D. Loughman, a colleague, a friend for over half a century, and a fellow Berkeley Ph.D.

FEEDBACK

The author would be pleased to receive comments about this book. Such comments may be sent by e-mail to mickeymck@prodigy.net.

TABLE OF CONTENTS

Page	Topic
1	Abstract
1	Introduction
3	Part 1: The Solution to Le Verrier's Problem
5	The Case of Mercury
7	Some Consequences of the Exact Solution
8	The Correct Form of Newton's Law of Gravity
9	Part 2: Gravity in the Solar System
11	Part 3: Derivation of Equation (1-1)
11	Differential Equations of Motion
13	*Vis Viva* Integral
14	The Main Differential Equation
15	Approximation 1
16	Approximation 2
19	Reference

TABLES

6	Table 1-1: Perihelion Precessions Corresponding to the Observed Perihelion Precession of Mercury
6	Table 1-2: Perihelion Precessions Calculated by General Relativity

Abstract

Einstein's theory of general relativity provided an answer to a long-standing physical problem: the precession of the perihelion of Mercury's orbit. That was a remarkable accomplishment. However, a hundred years later, an exact answer to that problem has been developed. The agreement between the two answers is close, but not exact. The difference shows that general relativity is not an exact theory, but only an approximation. As a consequence, the validity of deductions from the equations of general relativity cannot be assumed.

Introduction

One of Einstein's objectives in the development of general relativity was to explain a problem stated in 1859 by Urbain Le Verrier. (Le Verrier, a French mathematician and astronomer, is well-known as the man who discovered the planet Neptune by mathematical calculations.) The problem was that the location of the perihelion of Mercury's orbit as calculated by Newton's law of gravity did not agree with observations—the perihelion seemed to precess.

To make general relativity provide accurate results, Einstein made a few assumptions, notably that spacetime is curved.

One purpose of this paper is to offer an exact solution for that problem. This solution uses classical celestial mechanics in a Euclidean universe. The solution yields a result which differs slightly from that of general relativity; hence the word approximation in the title of this book.

Thus the exact solution does not assume that spacetime is curved. It shows that ordinary space is completely satisfactory for explaining Le Verrier's problem.

Part 1: The Solution to Le Verrier's Problem

The exact solution is really quite simple. It is based on the idea that the planets and asteroids which orbit the Sun do not experience the gravitational attraction of the entire mass of the Sun. That being the case, these planets and asteroids sense an apparent center of mass in the Sun which is offset from the Sun's true center of mass.

This leads to the idea of a *gravitational offset*. A gravitational offset exists in a star such as the Sun when the following condition is found: The apparent center of mass of the star, as sensed by a planet, asteroid, or comet orbiting the star, is offset from the true center of mass of the star. The following diagram (not to scale) illustrates this concept.

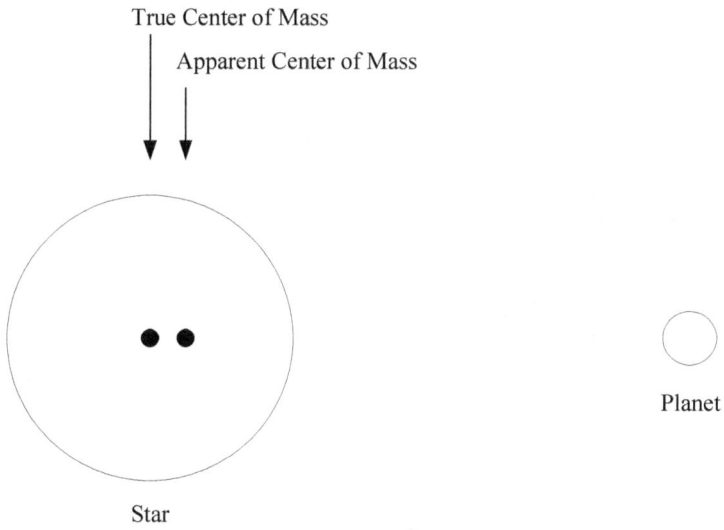

The gravitational offset is the distance between the true center of mass and the apparent center of mass of the star as sensed by the planet. It is the same for every planet, asteroid, and comet which orbits the star. Correct calculations for any object orbiting the star must apply Newton's law of gravity at the *apparent* center of mass of the star, not at the true center of mass.

3

If a gravitational offset exists in a star, then the observed mass of the star (its *gravitational mass*, found by noting its gravitational effect on its planets and asteroids) is less than the true mass of the star. This is because the location of the apparent center of mass implies that more mass is on one side of the star (the right side in the above diagram) than is on the other side.

In Part 3 of this book, classical celestial mechanics is used in Euclidean space to derive the orbit of a planet around a star having a gravitational offset. This derivation results in the following formula, which gives a direct relation between the amount of the periastron precession (or perihelion precession in the case of the Sun) and the size q of the gravitational offset.

(1-1)   Precession per orbit

$$= 360 \times 3600 \left(\frac{1}{\sqrt{1-2Lq}} - 1 \right) \text{ seconds of arc.}$$

In this equation, the parameter L is defined as

(1-2) $$L = \frac{G(M_1 + M_2)T^2}{4\pi^2 a^4 (1-e^2)},$$

where

    G is Newton's gravitational constant ($m^3 kg^{-1} s^{-2}$),
    M_1 is the apparent mass of the star (*kg*),
    M_2 is the mass of the planet (*kg*),
    T is the time of one orbit (*s*),
    a is the length of the semimajor axis of the orbit (*m*),
    e is the eccentricity of the orbit.

The precession per orbit can be converted into precession per Julian century (the units commonly used for this quantity) by multiplying it by the seconds per Julian century, and dividing by the seconds per orbit. The latter quantity is usually given in days, so the appropriate conversion factor is

(1-3) $\quad$ 3,155,814,977 / ((days per orbit) × 24 × 3,600)

where

> 3,155,814,977 is the number of seconds per Julian century.

Note that equation (1-1) shows that there can be no periastron precessions for objects orbiting a star if the star has a zero gravitational offset.

The Case of Mercury

The observed precession of Mercury's perihelion is 42.4446 arc-seconds per Julian century. From this and the orbital values for Mercury, it is simple to calculate the gravitational offset of the Sun. The necessary values for this calculation are

$G \quad 6.67384 \times 10^{-11} \ m^3 kg^{-1} s^{-2}$,
$M_1 \quad 1.9884 \times 10^{30} \ kg$,
$M_2 \quad 3.3011 \times 10^{23} \ kg$,
$T \quad 87.9691 \times 24 \times 3{,}600 \ s$,
$a \quad 57{,}909{,}050{,}000 \ m$,
$e \quad 0.205\ 630$.

Using these values in equation (1-1), with formula (1-3), gives the result $q = 4{,}375 \ m$. With this value, it is straightforward to calculate the precession of the perihelion of any object orbiting the Sun. Table 1-1 below shows some of those values.

Table 1-1
Perihelion Precessions Corresponding to the Observed Perihelion Precession of Mercury

Planet or Asteroid	Precession $\left(\text{arc sec}\left(\text{J-cent}\right)^{-1}\right)$
Mercury	42.4446
Venus	8.517349
Earth	3.790797
Mars	1.334068
1566 Icarus	9.931404
Jupiter	0.061486
Saturn	0.013299
Neptune	0.000761

For comparison, Table 1-2 shows the values yielded by general relativity

Table 1-2
Perihelion Precessions Calculated by General Relativity

Planet or Asteroid	GR-Calculated Precession $\left(\text{arc sec}\left(\text{J-cent}\right)^{-1}\right)$
Mercury	42.98
Venus	8.62
Earth	3.84
Mars	1.35
1566 Icarus	10.05
Jupiter	0.0623
Saturn	0.0137
Neptune	0.0008

The values in Table 1-2 are typically about 1.2% larger than the ones in Table 1-1. These values imply that the gravitational offset in the Sun is 4,429 meters. This shows how accurate general relativity is in this case.

Some Consequences of the Exact Solution

The exact solution used Newton's law of gravity to derive equation (1-1). This implies that when Newton's law is used correctly (that is, on the apparent center of mass of the Sun), the result is the observed behavior of the planets and asteroids orbiting the Sun. Thus Le Verrier's assertion that there is a flaw in Newton's law is incorrect; the flaw was in the way in which he used that law. (Of course, he had no way of knowing that. And his calculations were quite precise.)

An implication of the exact solution is that the assumptions used by Einstein to develop general relativity are only building blocks for the theory, but are not generally true.

Another consequence is that in order for a gravitational offset to exist, it must be that a star with a gravitational offset **blocks gravity**. The way in which that comes about is discussed in detail in the reference, but is quite simple in concept; the following paragraph briefly presents the idea.

A primitive particle (electron or proton) emits a signal at the rate of its frequency (cycles per second). This signal can interact with another electron or proton, as well as with another such signal. (The latter interaction is responsible for gravity; see the reference.) Evidence shows that the signal can interact with many other such signals on one cycle, but it can only interact with one other electron or proton on that cycle.

If a particle exists near the surface of the Sun, then when it emits a signal in the direction of the opposite side of the Sun, it has an extraordinary number of possible particles with which to interact as it passes through the Sun. In fact, there is a 100% chance that it will interact with an electron or proton before reaching the other side of the Sun. Consequently, it will not emerge from the Sun, and thus cannot interact gravitationally with a planet on the other side of the Sun. Thus, the gravitational effect of that particle is blocked by the mass of the Sun.

The Correct Form of Newton's Law of Gravity

Newton's law, when applied to two celestial bodies, is correct when the usual formula:

(1-4)
$$F = \frac{Gm_1 m_2}{r^2},$$

is interpreted in the following way:

- F is the attractive force between the two bodies,
- G is Newton's gravitational constant,
- m_1 is the gravitational mass of the first body,
- m_2 is the gravitational mass of the second body,
- r is the distance between the true centers of mass of the two bodies, minus the sum of the gravitational offsets of the two bodies.

Thus the force is applied at the apparent center of mass of each body.

Part 2: Gravity in the Solar System

The Sun is the only object in the solar system which has a gravitational offset. However, the planets themselves actually absorb some of the Sun's gravity. Thus when the moon of a planet lies in the planet's shadow, it will experience less gravity from the Sun than when not in the shadow.

So calculations of gravitational effects in the solar system fall into three categories. The first is calculations involving two bodies neither of which is the Sun. In this case, classical calculations using Newton's law of gravity yield correct results.

The second category is calculations between the Sun and another body. Here, Newton's law must be applied as interpreted in equation (1-4).

The third category is calculations of the orbit of a moon around its planet. In this case, it is appropriate to take into account the effect of the lessening of Sun's gravity on the moon as it passes through the shadow of the Earth. At present, there is no information available on the amount of this lessening.

Part 3: Derivation of Equation (1-1)

Differential Equations of Motion

The key idea which explains the shift in the perihelia of planets is this: When two bodies are in orbit around each other, two of the things which determine their orbits are not their centers of mass, but instead their *apparent* centers of mass. They sense each other's presence gravitationally, but each senses the center of mass of the other at a point possibly offset from its true center of mass. These offsets determine their mutual orbits.

Suppose two objects of gravitational masses M_1 and M_2 move in orbits around each other. Denote the positions of their centers of mass by the vectors $\mathbf{P}_1(t)$ and $\mathbf{P}_2(t)$. Let $\mathbf{P} = \mathbf{P}_2 - \mathbf{P}_1$, let $p = |\mathbf{P}|$, and let q_1 and q_2 be the sizes of the gravitational offsets of the two objects. Let $\mathbf{S}_1$ be the point between $\mathbf{P}_1(t)$ and $\mathbf{P}_2(t)$ which is a distance of q_1 from $\mathbf{P}_1(t)$, and let $\mathbf{S}_2$ be the point between $\mathbf{P}_1(t)$ and $\mathbf{P}_2(t)$ which is a distance of q_2 from $\mathbf{P}_2(t)$.

These assumptions are made:

A3.1   Each of the two objects acts as a rigid body.

A3.2 $\mathbf{P}_1$, $\mathbf{S}_1$, $\mathbf{S}_2$, and $\mathbf{P}_2$ lie on a straight line in that order.

A3.3   The offset distances q_i do not change with time.

Let $q = q_1 + q_2$.

The gravitational attraction between the two objects is applied at the two points $\mathbf{S}_1$ and $\mathbf{S}_2$. By the above assumptions, the distance between them is equal to $p - q$. Let $\mathbf{F}_i$ denote the gravitational force on object i; then (by analogy with Newton's theory)

$$\mathbf{F}_1 = \frac{GM_1M_2}{(p-q)^2}\frac{\mathbf{P}}{p}$$

and $\mathbf{F}_2 = -\mathbf{F}_1$.

By the first and second assumptions, the forces act on the gravitational masses, and the force vectors are applied at the apparent centers of mass. (One result is that the force on the apparent center of mass of the Sun contributes to the rotation of the Sun.) Hence, the differential equations of motion are

(3-1) $\qquad M_i \mathbf{P}_i'' = \mathbf{F}_i$,

where primes denote time derivatives. Since $\mathbf{F}_2 = -\mathbf{F}_1$,

$$M_1\mathbf{P}_1'' + M_2\mathbf{P}_2'' = 0,$$

and so the center of mass of the entire system:

$$\frac{M_1\mathbf{P}_1 + M_2\mathbf{P}_2}{M_1 + M_2}$$

moves in a straight line.

Now consider the differential equation for $\mathbf{P}$. Since $\mathbf{P} = \mathbf{P}_2 - \mathbf{P}_1$, it follows from equation (3-1) that

(3-2) $\qquad \mathbf{P}'' = \mathbf{P}_2'' - \mathbf{P}_1'' = \dfrac{\mathbf{F}_2}{M_2} - \dfrac{\mathbf{F}_1}{M_1} = \dfrac{-K\mathbf{P}}{p(p-q)^2}$,

where

$$K = G(M_1 + M_2).$$

Taking the vector product of $\mathbf{P}$ with the first and last sides of equation (3-2), it is found that $\mathbf{P} \times \mathbf{P}'' = 0$, which can be integrated directly to yield the fact that $\mathbf{P} \times \mathbf{P}'$ equals a constant vector. Thus, the orbits lie in a plane perpendicular to that vector. Take that plane as the *x-y* plane, and let the origin be at P_1.

Vis Viva Integral

Put equation (3-2) in rectangular coordinates:

$$x'' = \frac{-Kx}{p(p-q)^2},$$

$$y'' = \frac{-Ky}{p(p-q)^2}.$$

Multiply the first equation by $2x'$, the second by $2y'$, and add:

(3-3) $$2x'x'' + 2y'y'' = \frac{-2K(xx' + yy')}{p(p-q)^2}.$$

But by definition, the square of the velocity v is given by

$$v^2 = x'^2 + y'^2,$$

and so the left side of (3-3) equals $(v^2)'$. Also, $p = \sqrt{x^2 + y^2}$ and q is a constant. Hence,

$$(p-q)' = p' = \frac{2xx' + 2yy'}{2\sqrt{x^2 + y^2}} = \frac{xx' + yy'}{p},$$

and so the right side of (3-3) is equal to

$$\frac{-2K(p-q)'}{(p-q)^2}.$$

Thus, equation (3-3) can be integrated to obtain

(3-4) $$v^2 = \frac{2K}{p-q} + c_1$$

for some constant c_1.

The Main Differential Equation

Using polar coordinates, set

$$\mathbf{P} = p(\cos\theta, \sin\theta).$$

Then the second derivative is

$$\mathbf{P}'' = (p'' - p\theta'^2)(\cos\theta, \sin\theta) \\ + (2p'\theta' + p\theta'')(-\sin\theta, \cos\theta).$$

and the differential equation (3-2) becomes

$$\left(p'' - p\theta'^2 + \frac{K}{(p-q)^2}\right)(\cos\theta, \sin\theta) \\ + (2p'\theta' + p\theta'')(-\sin\theta, \cos\theta) = 0.$$

Since $(\cos\theta, \sin\theta)$ and $(-\sin\theta, \cos\theta)$ are mutually perpendicular unit vectors, the above sum can vanish only if the coefficients of those vectors are each zero; thus

(3-5) $$p'' - p\theta'^2 + \frac{K}{(p-q)^2} = 0$$

and

$$2p'\theta' + p\theta'' = 0.$$

From the latter equation, it follows that

$$(p^2\theta')' = p(2p'\theta' + p\theta'') = 0,$$

and so for some constant h,

(3-6) $$p^2\theta' = h.$$

Using this fact in equation (3-5) yields

(3-7) $$p'' - \frac{h^2}{p^3} + \frac{K}{(p-q)^2} = 0.$$

Next, use the standard transformation $p = 1/u$, and let the angle θ replace time as the independent variable, with dots indicating derivatives with respect to θ. By the definition of u, along with equation (3-6), it follows that

$$p' = \frac{-u'}{u^2} = \frac{-\dot{u}\theta'}{u^2} = -\dot{u}p^2\theta' = -h\dot{u}.$$

One more use of equation (3-6) yields

$$p'' = -h\ddot{u}\theta' = -h^2 u^2 \ddot{u}.$$

Hence, rewriting equation (3-7) in terms of u rather than p and dividing the result by $-h^2 u^2$ yields

(3-8) $$\ddot{u} + u = \frac{L}{(1-qu)^2},$$

where

$$L = \frac{K}{h^2}.$$

Approximation 1

Equation (3-6) represents twice the rate of accumulation of area in polar coordinates. The integral of (3-6) over one orbit (say for $t = 0$ to T, where T is the time required for the orbit) yields twice the area contained within the orbit. Under the (usually very accurate) approximation that the orbit is an ellipse, that area equals πab, where a and b are the lengths of the semimajor and semiminor axes. Hence, the integral results in the equation

$$2\pi ab = Th,$$

so that

$$h = \frac{2\pi ab}{T} = \frac{2\pi a^2 \sqrt{1-e^2}}{T},$$

where e is the eccentricity of the ellipse. Consequently,

(3-9) $$L = \frac{K}{h^2} = \frac{G(M_1 + M_2)T^2}{4\pi^2 a^4 (1-e^2)}.$$

Approximation 2

Each offset q_i is a fraction of the radius of the corresponding star or planet, and so the sum of the offsets, q, is much smaller than the separation p between the two objects. Thus the quantity $q/p = qu$ is much smaller than 1, and so for all u of interest, the denominator on the right side of equation (3-8) can be expanded into a power series which converges rapidly. In fact, an excellent approximation is obtained by dropping all terms of order u^2 and higher. The result is the approximate equation

$$\ddot{u} + u \cong L(1 + 2qu),$$

or

$$\ddot{u} + (1 - 2Lq)u \cong L.$$

The nature of the solution to this equation depends on whether $2Lq$ is smaller or larger than 1. It is not correct to say that $2Lq$ is small merely because qu is small. However, for the solution to be at all meaningful, $2Lq$ must be less than 1; otherwise, the solution would not be periodic, but would either expand or contract exponentially, which is not of interest. Hence for the moment, assume that $2Lq$ is less than 1; it is easy to show that it is positive. This leads to the solution

$$u = A\cos\left(\sqrt{1-2Lq}\,\theta - \omega\right) + \frac{L}{1-2Lq}$$

for constants A and ω, as is easily verified by differentiation. From this, it is possible to solve for p:

(3-10) $$p = \frac{1}{A\cos\left(\sqrt{1-2Lq}\,\theta - \omega\right) + \dfrac{L}{1-2Lq}}.$$

For small values of $2Lq$, this equation is very nearly that of an ellipse.

Recall that p represents the distance of the true center mass of the second object from that of the first object in a coordinate system in which the origin is at the true center of mass of the first object. The minimum value of p occurs when the denominator in equation (3-10) reaches its maximum, and that happens when the cosine equals 1. Similarly, the maximum value of p occurs when the denominator reaches its minimum, and that happens when the cosine equals -1. Of course, this presumes that the denominator never vanishes for any value of θ. It will be seen below that this is the case.

For an ellipse, the separation p at the time of periastron is equal to $a(1-e)$, and at apastron, p equals $a(1+e)$. Using the fact that equation (3-10) is very nearly that of an ellipse, form the equations

$$A + \frac{L}{1-2Lq} = \frac{1}{a(1-e)},$$

$$-A + \frac{L}{1-2Lq} = \frac{1}{a(1+e)}.$$

These can be combined to yield

$$\frac{L}{1-2Lq} = \frac{1}{a(1-e^2)}$$

and

$$A = \frac{e}{a(1-e^2)}.$$

17

The interval between two periastra is found by taking two successive values of θ for which the argument of the cosine is equal to a multiple of 2π. Two such values are θ_1 and θ_2, where $\theta_1 < \theta_2$, and where

$$\sqrt{1-2Lq}\,\theta_1 - \omega = 0$$

and

$$\sqrt{1-2Lq}\,\theta_2 - \omega = 2\pi.$$

for some value ω.

The change in θ from one periastron to the next is thus

$$\theta_2 - \theta_1 = \frac{2\pi + \omega}{\sqrt{1-2Lq}} - \frac{\omega}{\sqrt{1-2Lq}}$$

$$= \frac{2\pi}{\sqrt{1-2Lq}}.$$

If this change were equal to 2π, there would be no shift in periastron. Otherwise, the shift in periastron is equal to the above amount minus 2π:

$$2\pi \left(\frac{1}{\sqrt{1-2Lq}} - 1 \right) \quad \text{in radians}$$

$$360 \left(\frac{1}{\sqrt{1-2Lq}} - 1 \right) \quad \text{in degrees}$$

(3-11) $\quad 360 \times 3600 \left(\frac{1}{\sqrt{1-2Lq}} - 1 \right) \quad \text{in arc seconds.}$

Reference

McKinney, Albert W. III, 2015, *Gravity*, CreateSpace Independent Publishing Platform.

www.ingramcontent.com/pod-product-compliance
Lightning Source LLC
Chambersburg PA
CBHW061240180526
45170CB00003B/1377